I. C. PEECH
The internal combustion, piezoelectric, electrochemical hypothesis of the Great Pyramid of Giza
Terry E. Lee

Terry E. Lee

Copyright © 2025 by Terry E. Lee

All rights reserved.

No portion of this book may be reproduced in any form without written permission from the publisher or author, except as permitted by U.S. copyright law.

The antique photos and illustrations 95 years or older are considered public domain.

Contents

Introduction   1

Acknowledgements

1. Chapter 1   5
   Old world high tech

2. Chapter 2   10
   Initialization

3. Chapter 3   18
   The Engine of the Great Pyramid
   I. C. Engine

4. Chapter 4   24
   The Voltage Source

5. Chapter 5   28
   Ionization and Townsend's Current Growth Equation

6. Chapter 6   30
   Electrolysis and the steam chamber or water to fuel converter
   The Steam chamber

7. Chapter 7   36
   Power output and the relieving chambers

8. Chapter 8   38
   The shoulders of giants

9. Chapter 9   44
   Iron plate and the Signal Beacon

10. Chapter 10   47
    Wireless receivers, user devices and the Wear & Tear evidence
    Wireless receivers and user devices

11. Chapter 11   56
    Volumes and values

12. Chapter 12   58
    Conclusion
    Print index

    Print index
    Print index
    Print index
    Print index
    Print index

13. Chapter 13   68
    Bonus pics
    NOTES:
    NOTES:

Fullpage image   77

Introduction
Acknowledgements

Together we stand on the shoulders of the giants who came before us, too many to mention here but a special Thank You to the genius of Nikola Tesla. His vision and pioneering work gave rise to our modern world and continues to point the way to our future developments.

Much gratitude to our modern community of content creators for sharing and discussing their knowledge and ideas about so many subjects. Today's popular platforms and new generation of scientists and researchers color the modern era and makes way for a new chapter in the advancement of human knowledge. These content creators deserve our thanks for their efforts and contributions for the benefit of people everywhere.

Also a special thank you to the hard work of prominent authors and researchers, their efforts have helped open our eyes to the technology of a previous civilization. Their contributions continue to resonate and inspire the next generation of researchers like ourselves.

Hypothesis: A proposed explanation made on the basis of limited evidence as a starting point for further investigation.

Theorem: A general proposition not self evident but proved by a chain of reasoning; A truth established by means of accepted truths.

Introduction

The Internal Combustion, Piezoelectric, Electrochemical Hypothesis of the Great Pyramid of Giza

From a layman's perspective in the study of this building, one can't help but notice that the Queen's chamber seems designed to hold water and the Queen's chamber Niche just has a look, like it contained something hot just above the water line. Having said that, the hypothesis and theory of operation arises from the appearance of wear and tear evidence present on key features of the Great Pyramid, practical reasoning and the implications of physics.

The Great Pyramid is a piezoelectric power generator, Hydrogen / Oxygen fueled, internal combustion. The Queen's chamber held water and liberated fuel through pyrolysis and electrolysis, contained a heat source in the Niche and the "air shafts" are high voltage conductors. The King's chamber is made of piezoelectric material and the sarcophagus contained the igniter.

Some tell tale signs of wear and tear would include the Queen's chamber being found encrusted with some type of salt deposits.

The substantial erosion and erosion patterns present on the limestone Great Step.

The King's chamber's Antechamber bears a thermal signature of having been scrubbed with hot, high velocity gasses.

The severe erosion all around the top of the sarcophagus and the appearance of worn electrodes at the top of the Queen's chamber "air shafts".

SYNOPSIS

This text presents an argument for the presence of an advanced technological civilization from our distant past. Perhaps when we see the pyramids we are looking at some of the few remains from the cataclysm upon the earth known as the *Younger Dryas event*,[1] about 12,000 years ago.

Basic physics and insights from a recently declassified book from 1963,[2,3,4] suggests that when the polar regions become heavy with ice, the spinning ball of the earth {or oblique spheroid if one prefers} can roll onto it's side and the poles become the equator... until enough of the ice caps melt and the earth flips back to an upright position. Of course this would result in large scale tsunami's and cataclysm causing widespread destruction and upheaval around the planet, a rapid rise in sea levels and a drop in global temperatures.

The theory of operation for the Great Pyramid is presented in sequence from water fill and gas purge to system operation and power output. An effort is made to use plain language and basic concepts of physics, common knowledge or at least commonly available knowledge in the twenty first century.

So please grab a favorite beverage and return your seat to a reclined position while we begin.

***

[1] www.wikipedia.com "Younger Dryas"

[2] www.wikipedia.com Chan Thomas, author "The Adam and Eve Story" 1963 Bengal Tiger Press

[3] www.cia.gov Chan Thomas "Adam and Eve Story" declassified and sanitized 2013

[4] www.youtube.com The Why Files 2023 "CIA classified book about the pole shift, mass extinctions and the true Adam and Eve story"

Chapter 1

Old world high tech

The Great Pyramid is only one example of advanced energy technologies present in our distant past, albeit not the most readily apparent. Perhaps the most visible, yet hidden in plain sight example of high tech from the past would be the ancient remains of what appears to be some type of resonant cavity device the size of a house, located atop the hill at Sacsayhuaman, just outside the city of Cusco, Peru.

The ancient builders of Cusco, Sacsayhuaman, Ollantaytambo and Machu Picchu clearly utilized an advanced stone working technology. An understanding of *this* device should tell us what we want to know about how the ancients were able to manipulate stone like it was putty. But that's a subject for another time as this book will focus solely on the operation of the Great Pyramid.

Great Pyramid internal cavities

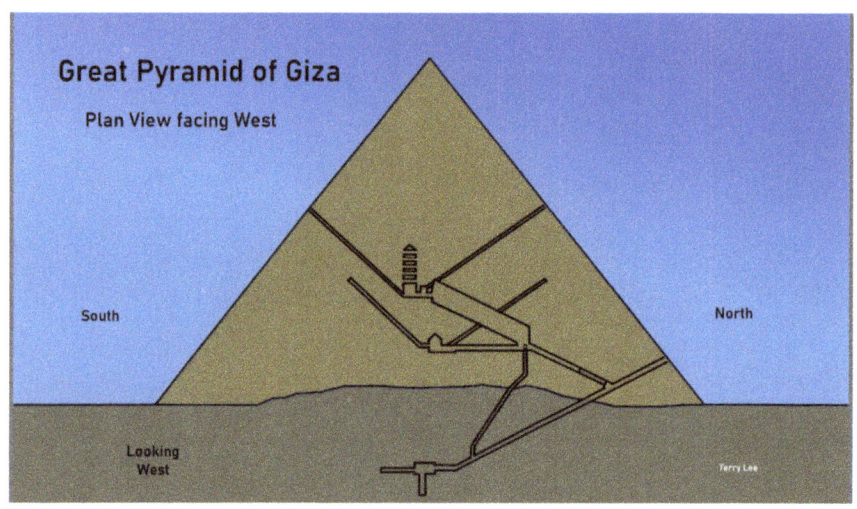

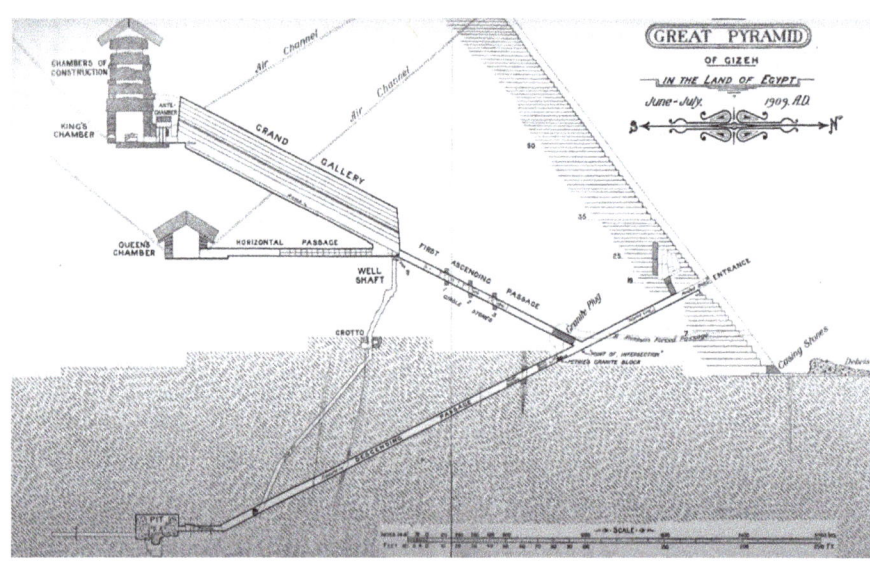

Impression of operation

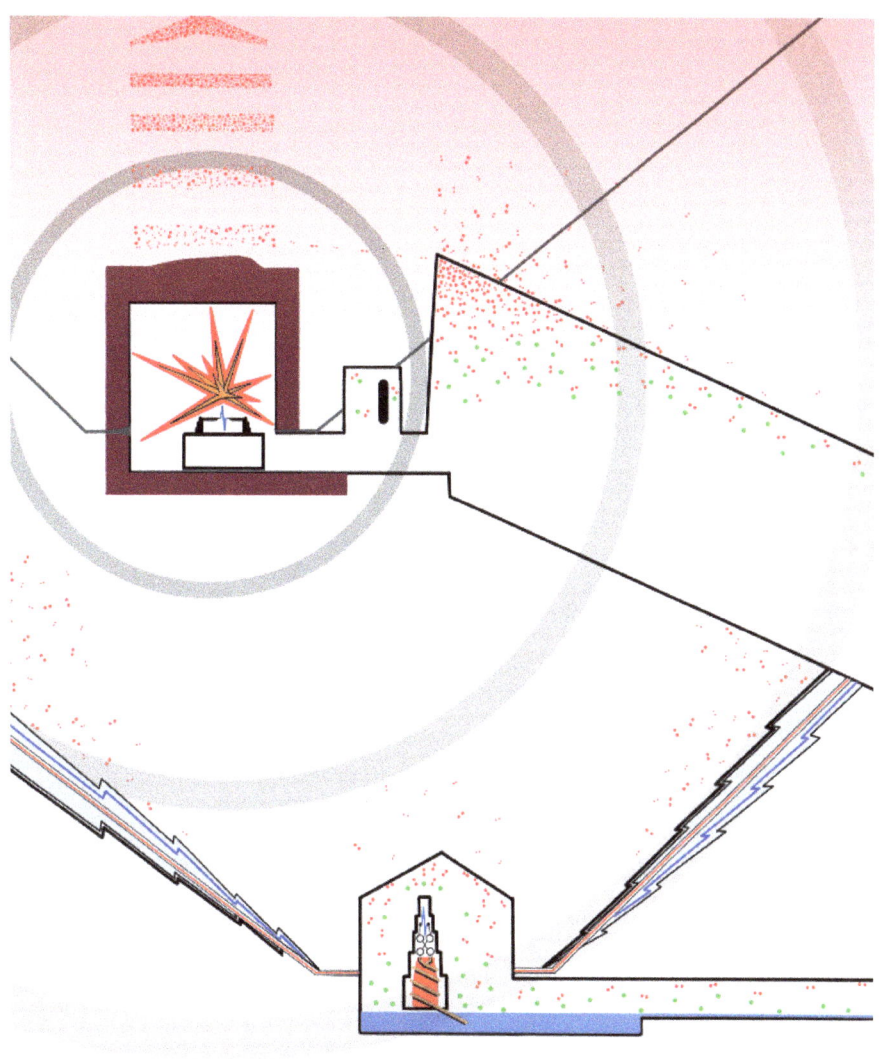

Terry E. Lee

Component identifier

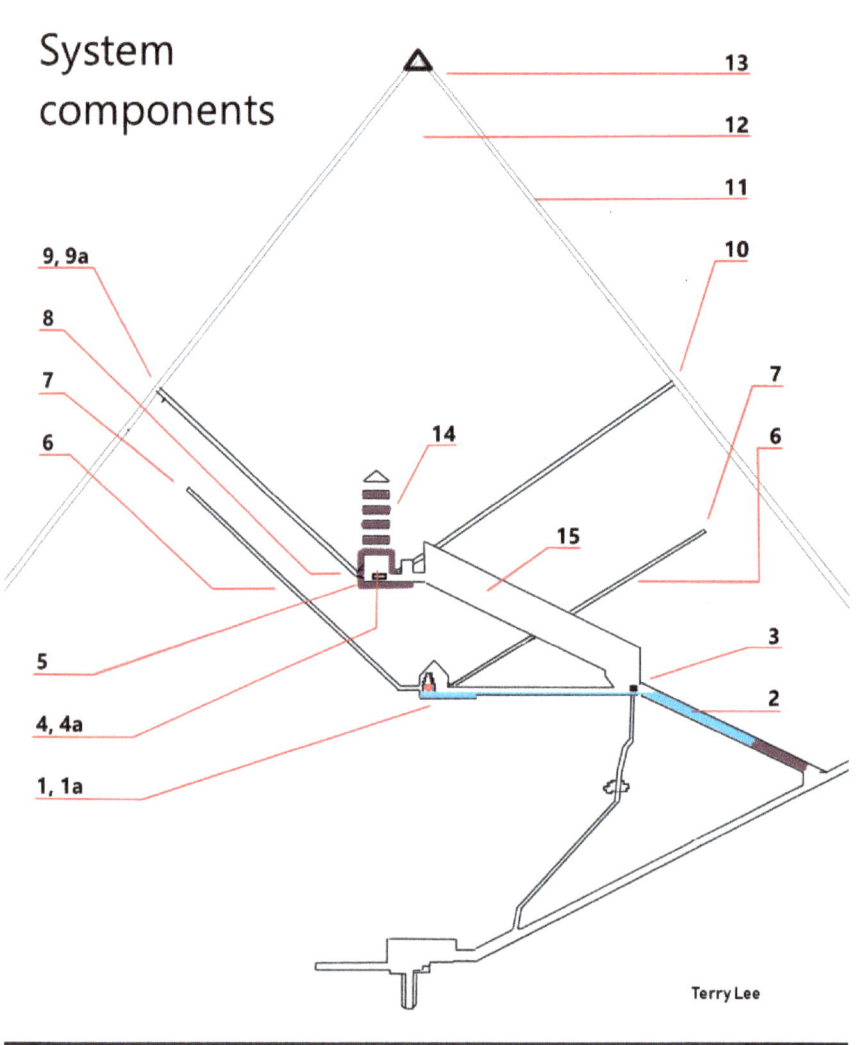

SYSTEM COMPONENTS

1 Steam chamber, water to fuel converter

1a Heat source, steam powered electrostatic generator

2 Water reservoir

3 Drain

4 Combustion chamber

4a Igniter

5 Piezoelectric blocks

6 High voltage conductors

7 Electrodes

8 Scoop

9 Signal beacon

9a Iron plate

10 Vent

11 Limestone insulator

12 Gas capacitance

13 Conductive final

14 Gas cavity amplifier

15 Tuning chamber

Chapter 2

Initialization

Once the construction and assembly of the system was complete, it seems reasonable that the start up process would begin by filling the system with water and turning on the heat source. Of course the heat source is missing, leaving much room for speculation about what exactly was there.

The hypothesis holds that the energy for the system was supplied from a heat source located in the Niche of the Queen's chamber.

For the purpose of presenting the theory of operation it seems prudent to describe the heat source as being compact and efficient, something akin to the SNAP reactor program developed for NASA in the 1960's to power their early satellites.[1,2] In addition, the space provided within the Niche would accommodate the hot end of the SNAP reactor unit just about right.

Filling the system with water

The liquid capacity of the Queen's chamber and the water reservoir appears to hold an estimated 80 cubic meters, or 80,000 liters of water. Filling the system via the King's chamber vent shaft would spill the water onto the floor and down to the Queen's chamber. Adding 80,000 liters of water would fill the Queen's chamber and the water reservoir up to the level of the drain, while the excess would drain into the subterranean chamber and out through the pit.

[1] www.wikipedia.com "Systems for Nuclear Auxiliary power"

[2] www.youtube.com Periscope Film 2015 "First nuclear reactor in space, SNAP 10A program 1965"

Queen's chamber

The heat source and production of Hydrogen / Oxygen gasses

With the system filled with water up to the level of the drain, it appears the water would come up to just below the level of the Niche. The original heat source is missing and while its operation is unknown, the hypothesis suggests that the heat source drew a stream of water from the pool and heated the water to a superheated steam. The steam was mixed with liquid water via a venturi tube and directed through a series of coil sets, know as *'Kelvin's thunderstorm generator'*.[1] This produces high voltage electrical arcs through the steam causing some of the water molecules to decompose through pyrolysis [2] and electrolysis,[3] producing Hydrogen and Oxygen gasses.

Hydrogen.[4] The lightest element, the tiniest atom and an ionizing gas

Once the system was initialized and gas production had begun, Hydrogen being the lightest element would tend to rise to the highest points and displace the heavier Nitrogen and Oxygen of the air. Hydrogen also being the tiniest atom would tend to leak through every possible crack or crevice while always spilling upward in the process. The Hydrogen presents a gaseous dielectric medium that serves as an electrical conductor at high voltages through primary ionization and acts as an amplifier at sufficiently high voltages through the avalanche effect of secondary ionization [5] as described in Townsend Discharge.[6]

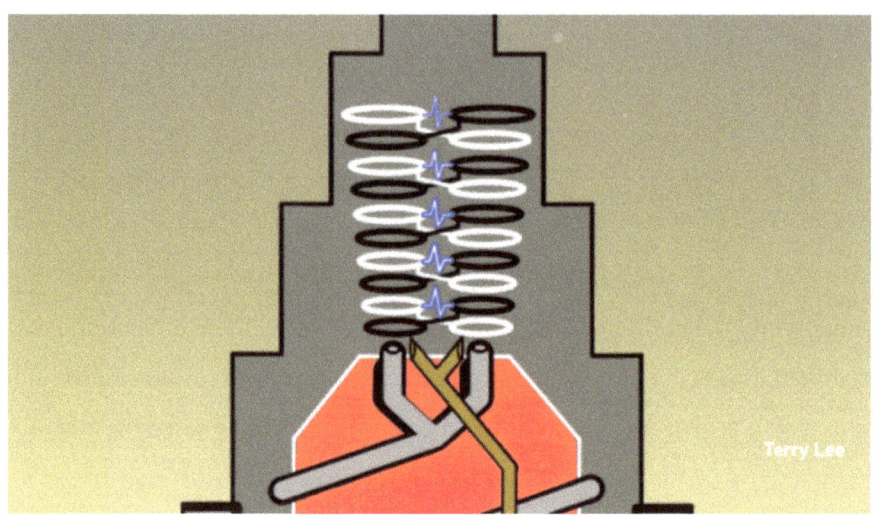

***

¹ www.wikipedia.com "Kelvin Water Dropper"

² www.wikipedia.com "Electrolysis"

³ www.wikipedia.com "Pyrolysis"

⁴ www.youtube.com   Thoisoi 2 2023 "Hydrogen, the lightest gas in the universe"

⁵ www.wikipedia.com "Electron Avalanche"

⁶ www.wikipedia.com "Townsend Discharge"

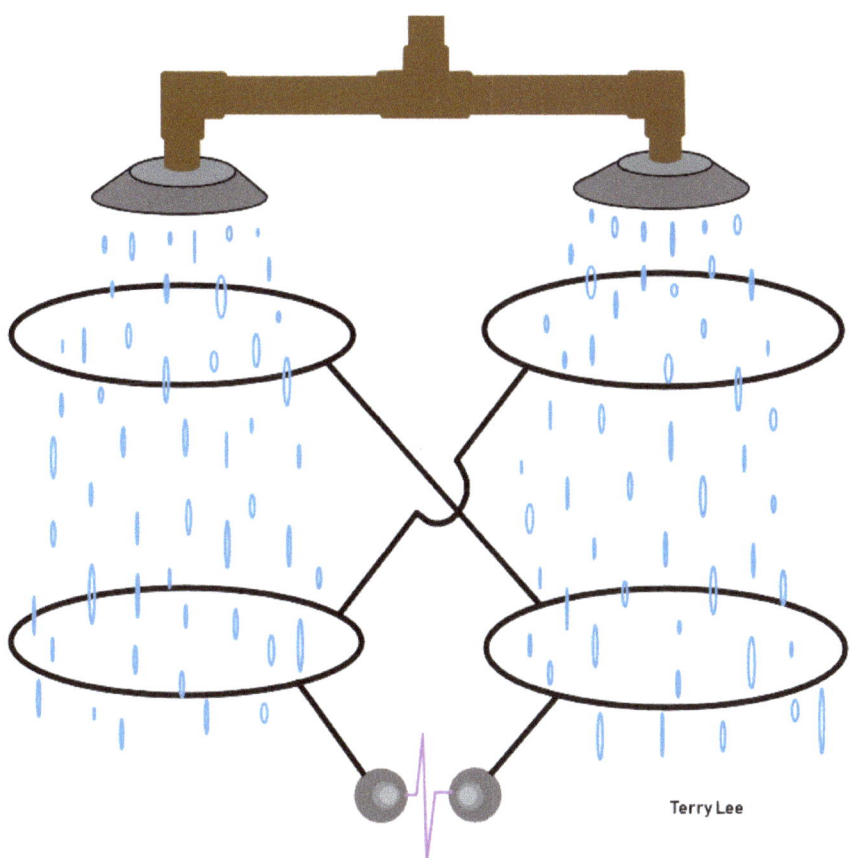

Kelvin's Water Dropper experiment

Also known as Kelvin's Rainstorm, or Lord Kelvin's Thunderstorm Generator. The experiment shows that dribbling water droplets through a set of cross connected coils will produce high voltage electrical arcs of 10kv - 20kv between the electrodes.

Gas Purge

Gas production would need to run for some period of time to achieve a partial Hydrogen saturation of the building and enable system start up. Once the gas production was running and the air was being purged from the building, the Hydrogen would tend to pool on the ceilings and fill the spaces from the top down while permeating the stonework through every available crack or crevice.

The Oxygen gas is heavier and would tend to pool at the lowest point where the drain is located. The drain provides a path for the heavier Oxygen to escape through the drain cavity and ultimately out the front door.

Given these conditions, the building would eventually become partially saturated with Hydrogen, the air purged out and the internal cavities filled with steam, water vapor and Hydrogen / Oxygen gasses. With the initialization process complete, the building purged and primed with H^2 gas, the system is ready for start up and operation.

A water vapor environment

It presents a twist on physics to consider humidity, or the water content of the air since all of the air would be purged from the building and only steam, water vapor and HHO fuel gasses would occupy the interior spaces of the building.

The temperature / pressure relationship with water and the behavior of steam would feature prominently in the operation of the combustion chamber. When the HHO fuel gas reacts and burns to form water in the combustion chamber, a shockwave is produced due to HHO having the highest specific impulse of all known fuels. The reaction is exothermic and the newly formed water molecules along with some of the surrounding water vapor would flash to steam, producing a sharp rise in combustion chamber pressure. Chamber pressure would then drop to a relative vacuum due to rapid cooling of the steam, drawing in a fresh gulp of water vapor and fuel gas mixture for the next combustion cycle.

The space at the top of the Grand Gallery provides a reservoir for the HHO fuel gas and is thought to work in tandem with the Antechamber to accomplish fuel metering.

Chapter 3
The Engine of the Great Pyramid

I. C. Engine

Tuning chamber

Since the system primarily operates as an internal combustion power plant, it seems prudent to point out here that the length of the Grand Gallery would serve as a tuning chamber. Somewhat similar to the purpose of a *"Stinger"* type exhaust system equipped on high performance two cycle dirt bike engines.

The shape of the Stinger exhaust is designed to reflect the pressure wave of the exhaust pulse back to the exhaust port.[1,2] The length of the Stinger exhaust is calibrated so that the reflected pressure wave arrives back at the exhaust port at the desired moment in time, corresponding to the peak power design speed of the engine. Similarly, the length of the Grand Gallery provides a clue to the system's frequency of operation, which works out to about 4 hertz, or 240 cycles per minute.

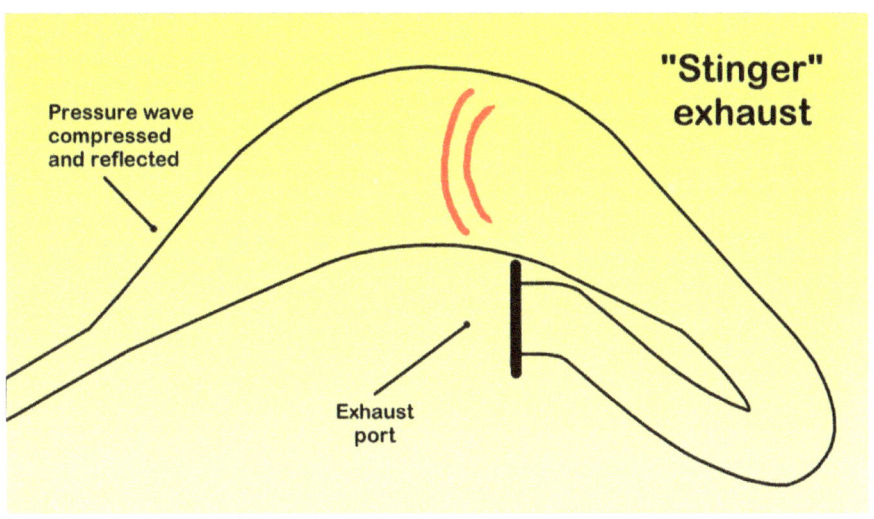

The Igniter

The igniter containment is located near the northwest corner of the combustion chamber, this location becomes significant when considering the propagation of the pressure wave within the combustion chamber. Since the nucleation point of the detonation occurs at the northwest corner of the chamber, the focal point would be located at the opposite southeast corner where the south wall shows extensive damage around the southern shaft.

[1] www.youtube.com driving 4 answers 2024 "How two stroke exhaust pipes really work"

[2] www.youtube.com Steps to Podium 2020 "How 2-stroke exhaust pipes work"

The igniter containment box is of course empty, again leaving much room for speculation about what exactly was there. Here the hypothesis suggests that the igniter consisted of some sort of capacitor containing a radioisotope { such as Radium 226 } that provided a sufficient rate of charge through the electron emission of Beta decay,[1,2] to continuously charge the capacitor and featured a discharge spark gap. The rate of charge, the capacitance of the capacitor and the distance of the spark gap calibrating the frequency of the spark output and the system's operation.

The people's choice award

[1] www.wikipedia.com "Beta decay"
[2] www.energy.gov U.S Dept. of Energy "DOE explains Beta decay"

The combustion chamber

The combustion chamber is constructed of quartz bearing, red granite blocks of rather large proportion and constitutes the system's voltage source. The quantity of piezoelectric crystal contained in the combustion chamber construction is unclear, as is the portion of the crystal that would contribute to the generation of an electromagnetic pulse.

What *is* clear is that the detonation of a stoichiometric Hydrogen / Oxygen fuel mixture in a chamber constructed of material loaded with piezoelectric crystal would generate a high voltage and an electromagnetic pulse of unknown magnitude.

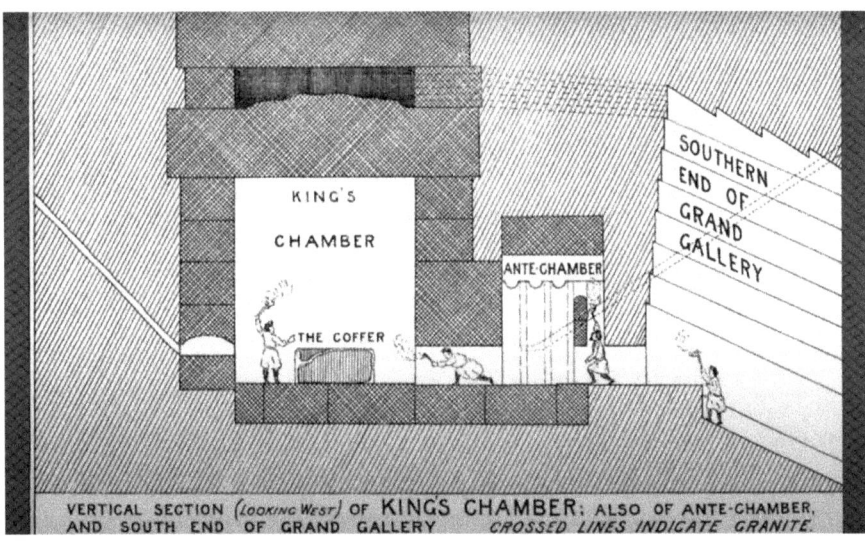

System operation

The length of the Grand Gallery provides a clue to the system's design speed or frequency of operation. Curiously enough, the speed of sound increases with higher temperature and humidity levels to about 440 meters per second at 100% humidity and 100^0C.[1] It's worth mentioning that the humidity level of the Grand Gallery would likely exceed 100% to some degree of supersaturation due to the continuous production of steam from the steam chamber.

Each combustion event in the combustion chamber would produce a pressure pulse that travels the length of the Grand Gallery and then reflects back to its source. Since the distance between the combustion chamber and the bottom of the gallery is just under 60 meters, the pressure pulse would take about 250 milliseconds to make the round trip suggesting an operating frequency of about 4 hertz. It would seem optimal for this time frame to coincide with the time it takes for combustion chamber pressure to drop to a relative vacuum after a combustion event due to rapid cooling of the steam, thus helping to fill and swirl the combustion chamber for the next combustion cycle.

[1] www.wikipedia.com "Speed of sound"

CHAPTER 3

Chapter 4

The Voltage Source

The quartz crystal contained within the red granite blocks of the combustion chamber serves as the system's voltage source via the piezoelectric effect.[1,2,3]

The Silicon atoms carry a small positive charge while the Oxygen atoms are slightly negative. The symmetric Silicon Oxygen lattice of the quartz crystal is electrically neutral until a shockwave or vibration causes a distortion of the crystal lattice. When the Silicon Oxygen lattice is compressed or distorted on the appropriate axis, the crystal becomes more positive at one end and more negative on the other end. This momentary polarization causes a potential difference across the crystal and a flow of electrical current from the negative side to the positive side.

Quartz crystal structure

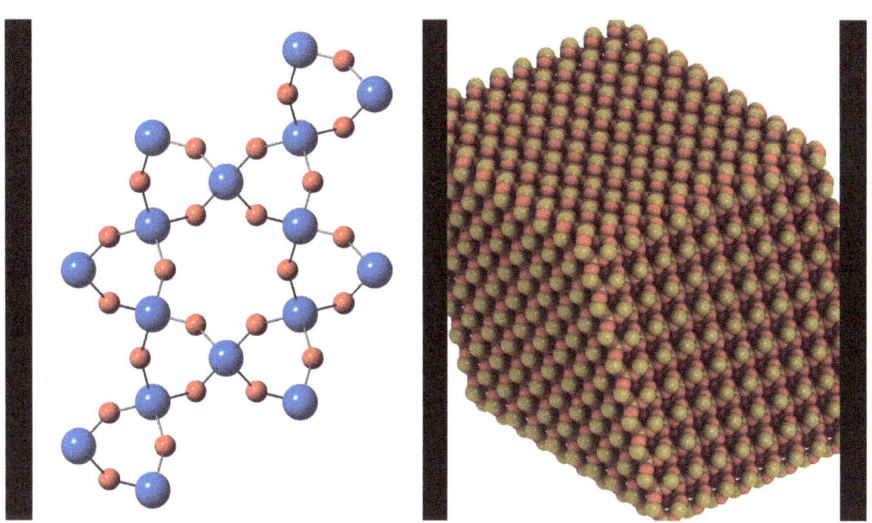

[1] www.wikipedia.com "Piezoelectricity"

[2] www.youtube.com Physics High 2017 "Piezoelectric effect explained"

[3] www.youtube.com Steve Mould 2019 "Why hitting crystals makes electricity"

Silicon / Oxygen lattice

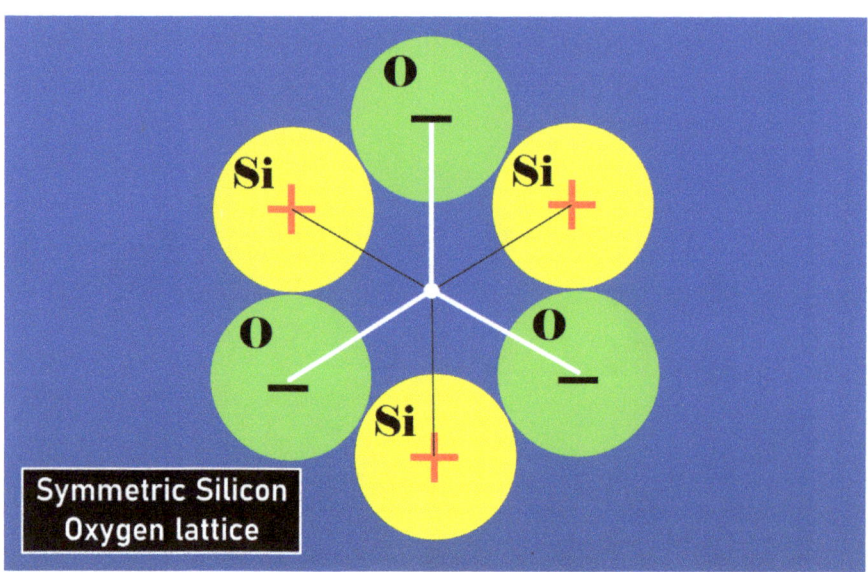

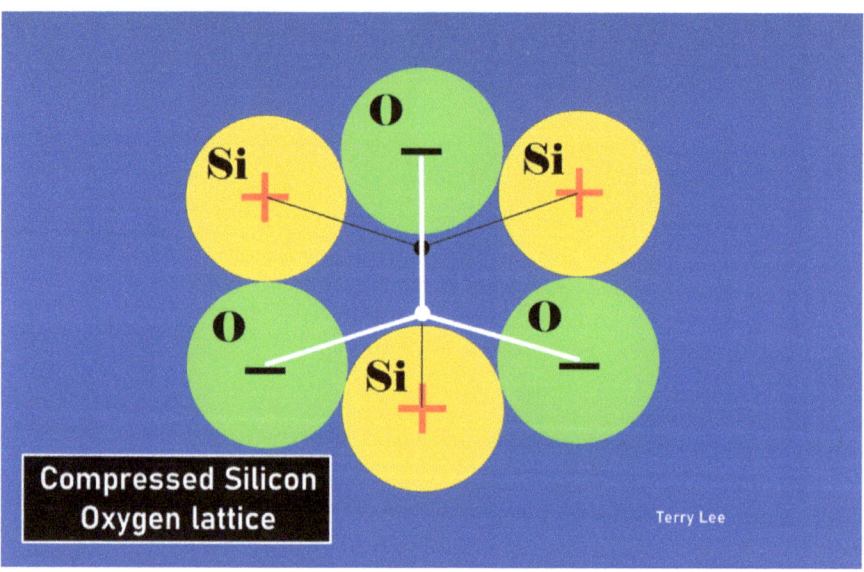

CHAPTER 4   27

Piezoelectric current

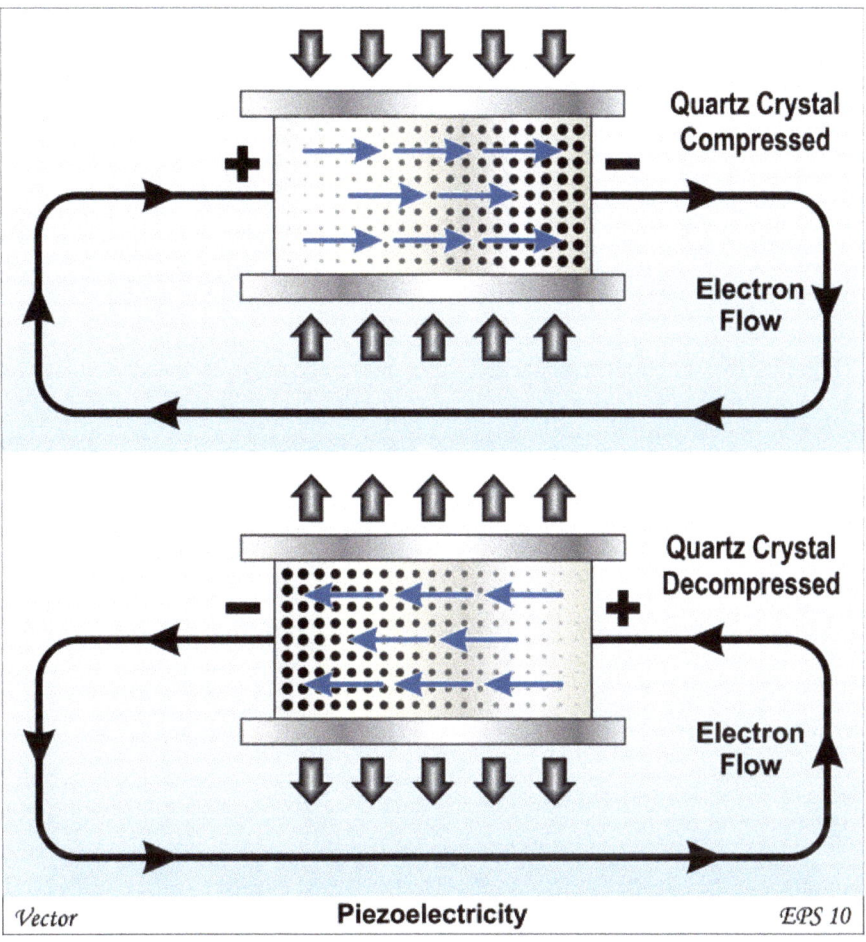

Chapter 5

Ionization and Townsend's Current Growth Equation

The processes of primary ionization and avalanche effect of secondary ionization in a gaseous dielectric medium are described in *Townsend's Theorem*.[1] There Townsend shows an avalanche effect where electrons and ion are multiplied exponentially at increasingly higher voltages with his current growth equation.[2,3]

Interested readers should of course review Townsend's theory for a proper description of primary vs secondary ionization and the current growth equation.

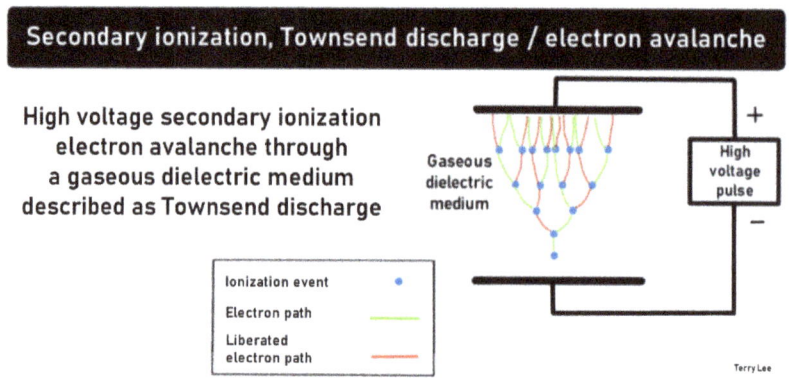

Ionization in Hydrogen gas [4]

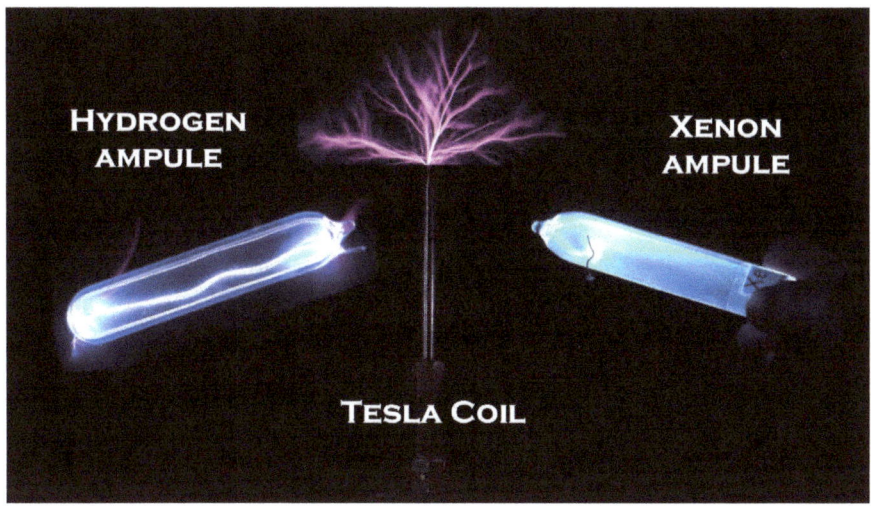

---

[1] www.youtube.com Sai Academy of engineers 2021 "Townsend's primary and secondary ionization coefficient"

[2] www.wikipedia.com "Current growth through secondary ionization"

[3] www.youtube.com Sai Academy of engineers 2021 "Townsend's current growth derivation"

[4] www.youtube.com Thoisoi 2 2023 "Hydrogen, the lightest gas in the universe"

Chapter 6
Electrolysis and the steam chamber or water to fuel converter

The Steam chamber

The high voltage conductor tubes of the steam chamber angle upwards through the building to the north and to the south of the chamber from the building's centerline. Both of the tubes terminate at a stone plate featuring a pair of metallic looking projections with the color of corroded Copper and the last few feet of the tubes are constructed of a different type of stonework than the rest of the tube, suspected to be the same low Magnesium Tura limestone used for the building's exterior casing.

The upward angle and the cavity space of the tubes present another place for the lightweight H^2 gas to be displaced into while it leaks through and permeates the stonework at every opportunity on its path upward. The tubes become conductive when the high voltage EMP [1] sweeps past and ionizes the gaseous dielectric medium contained within the tubes and upwardly surrounding stonework.

CHAPTER 6

During a combustion event, a high voltage electromagnetic pulse is generated and propagates from the voltage source in all directions. The hypothesis suggests that the voltage is sufficient to produce the avalanche effect of secondary ionization, the EMP carries electrons and grows in strength as it sweeps through the gas content of the building like a wave.

The voltage source is offset from the centerline of the building to the south side and the southern shaft is somewhat closer to the voltage source than the northern shaft. As the wave of electrons propagates through the building it reaches the southern shaft before it reaches the northern shaft.

***

[1] www.wikipedia.com "Electromagnetic pulse {EMP} Also referred to as a transient electromagnetic disturbance"

Steam chamber, water to fuel converter

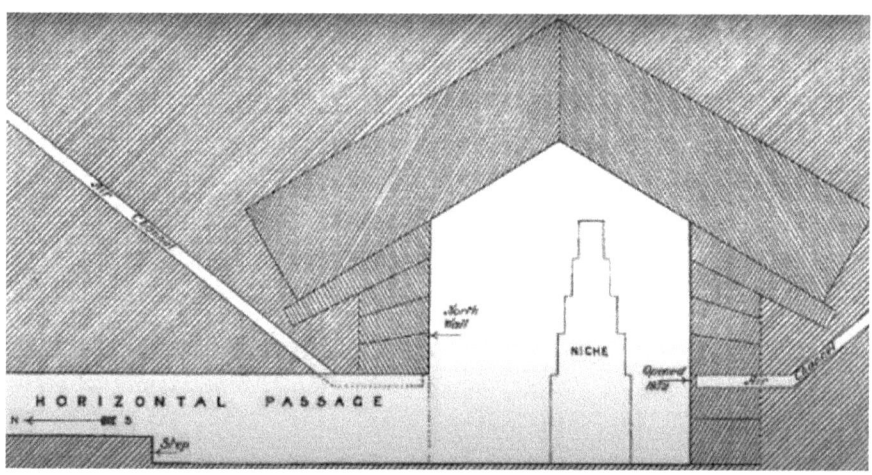

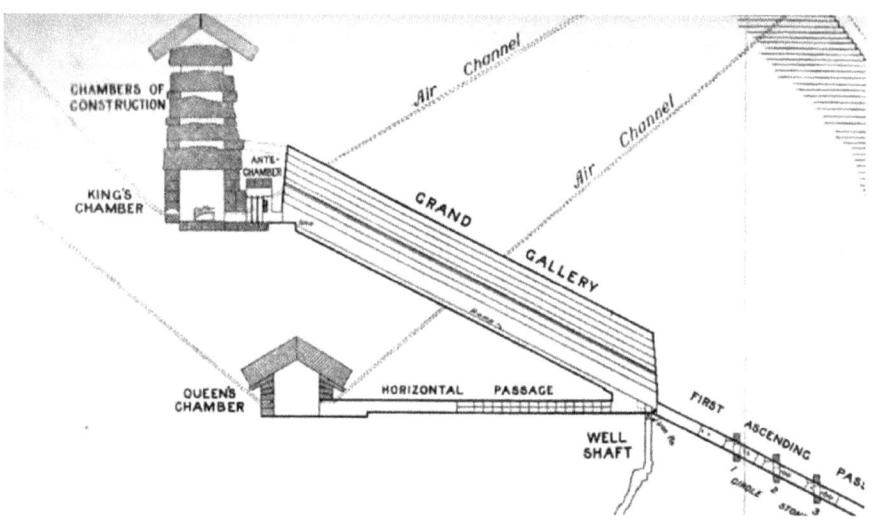

The avalanche effect of secondary ionization strips electrons which are carried along with the wave per Townsend's current growth equation, leaving positively charged ions in its wake. This causes the southern shaft to become positively charged at the same moment the wave of electrons reaches the northern shaft, creating a large potential difference between the shafts and across the steam chamber.

The hypothesis suggests that the potential difference is sufficient for the abundance of electrons to jump the 5.7 meter gap across the chamber to the positively charged south side. The resulting electrical arc through the steam within the chamber producing HHO fuel through pyrolysis and electrolysis, perhaps best described as plasma HHO production.

The resonant property of the King's chamber also provides a clue to the frequency of electrical arcs produced through the steam chamber. Given the dimensions of the combustion chamber and the speed of sound at 100^0C and 100% humidity, the acoustic frequency of the chamber works out to 72 hertz north and south while the east – west component runs at 36 hertz. The velocity of the P wave through granite is roughly 6 kilometers per second, which suggests a 3 kilohertz vibration frequency for the walls of the chamber at an estimated one meter thick.

It's unknown what feature inspired the Dixon party to chisel out the wall where the tubes are located on both sides of the Queen's chamber in 1872. The only hint we have is what we see at the top end of both the northern shaft and the southern shaft.

[1] www.youtube.com Ancient Architects 2021 "EXCLUSIVE: First look inside the Queen's chamber northern shaft"

CHAPTER 6

QUEEN'S CHAMBER, 1904

Chapter 7

Power output and the relieving chambers

The series of cavities and complex of shaped beams known as the relieving chambers would appear to serve two functions for the system. It's of particular interest that each one of the beams in the complex is made flat on the bottom and sides, yet each beam features a unique profile on their top surface. Taken together as a whole functioning component, the apparently random undulations of the beams would serve to scatter or diffuse energy moving vertically through the complex. More importantly, these chambers present another upward cavity for the H^2 gas to be displaced into and collected. This places a series of gas filled cavities directly between the voltage source and the power output.

The hypothesis suggests that when an electromagnetic pulse propagates from the voltage source in all directions, the voltage is sufficient to produce the avalanche effect of secondary ionization through the gaseous dielectric medium. As the EMP propagates through the building it carries a wave of electrons that grows in strength. The upwardly moving portion of the wave is amplified via the gas filled cavities and funneled to the pyramidion where it yields from the pointy spot as an electrical arc of unknown magnitude.

CHAPTER 7

Gas cavity amplifier

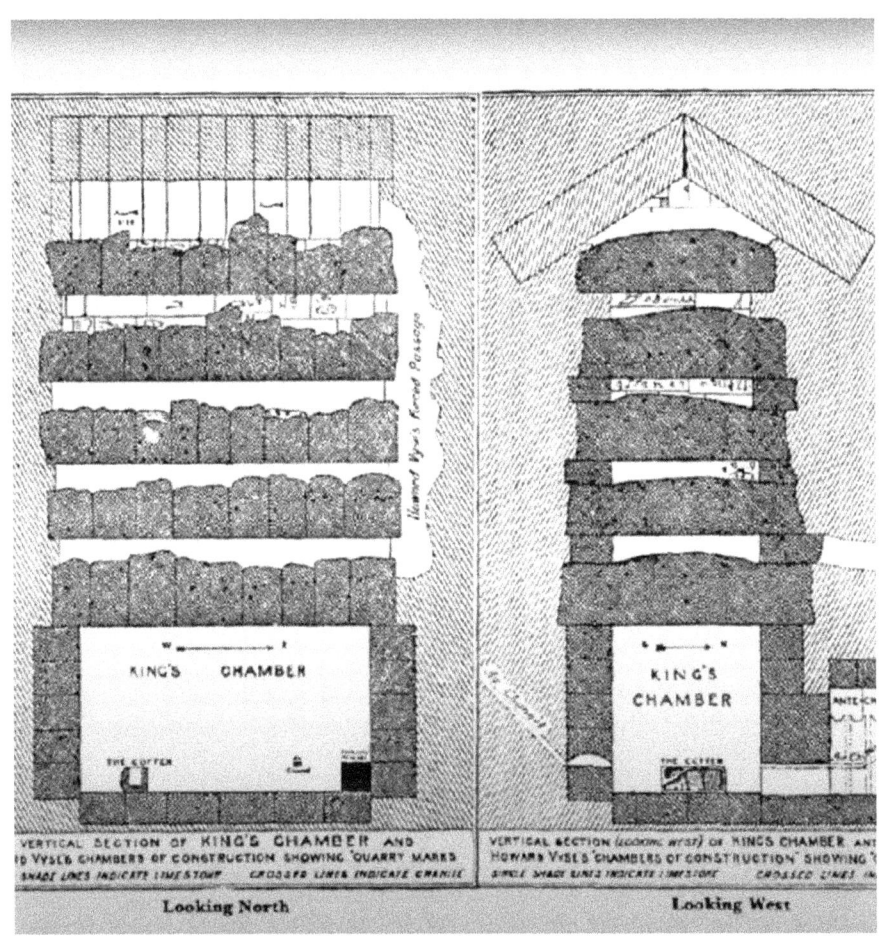

Chapter 8

The shoulders of giants

Nikola Tesla, the Wardenclyffe Tower and the Tesla Coil

We know that Nikola Tesla wanted to operate his Wardenclyffe tower at 8 hertz, he believed that his wireless power transmitter would achieve resonance at 8 hertz due to the speed of light and the circumference of the earth. This would allow people everywhere to attenuate the wireless power with an antenna he described simply as a metal plate. He also said that one could operate their lamp by inserting a wire into the ground, that is all.

In the classic Tesla Coil design, a capacitor is charged to a high voltage until the space across the spark gap becomes ionized allowing the charge to jump the gap and flow through the pancake shaped primary coil. This produces an electromagnetic pulse that induces a current and voltage into the secondary winding up the central tower to the toroid shaped capacitor atop the tower.

The capacitor is toroid shaped to add capacitance and avoid pointy spots where charges will concentrate causing electrical arcs to yield from the pointy spot of the capacitor. Both the primary side and the secondary side are tuned resonant circuits that oscillate at a high frequency.

Interested readers are recommended to review the excellent presentation suggested below for a thorough description and explanation of the spark gap Tesla Coil's operation in this six part, do it yourself series.[1]

***

[1] www.youtube.com Diode Gone Wild 2020 "How to build a Spark Gap Tesla Coil"

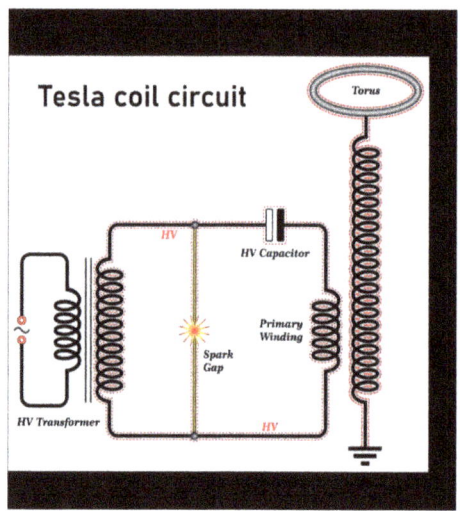

Wardenclyffe Tower

CHAPTER 8

Credit to prominent authors and researchers

Much gratitude to the dedicated researchers and authors of recent decades. Their efforts and hard work has helped point the way to an understanding of the Great Pyramid. Their works carefully document aspects of the Great Pyramid including the King's chamber's southern shaft and its funnel shaped opening.[1] One of the most prominent theories suggests that the southern shaft constitutes a microwave guide.

Radiation and the x-ray tube -vs- the resonant cavity magnetron

The physics of these two devices provide a relevant example of how radiation would be produced in the King's chamber's southern shaft.

In the x-ray tube, electrons are liberated from the heated cathode due to thermionic emission.[2] The negatively charged electrons accelerate as they are drawn toward the high positive voltage anode and radiation is produced when the high velocity electrons collide with the high voltage anode.

Sort of the opposite is true in the resonant cavity magnetron where the heated cathode is driven to a high negative voltage, the electrons are spun by a magnetic field and accelerate to a high velocity as they are drawn toward the relatively positive resonant cavity anode which is maintained at a ground potential.[3]

X-ray tube and resonant cavity magnetron

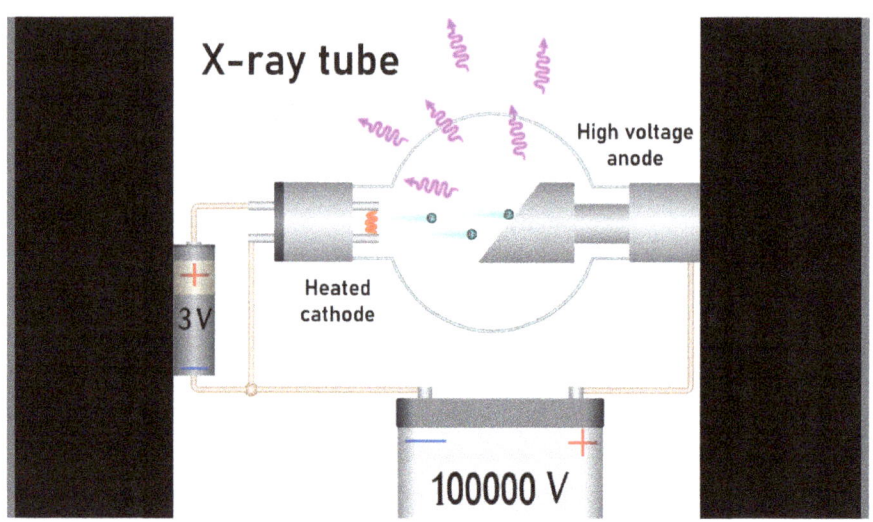

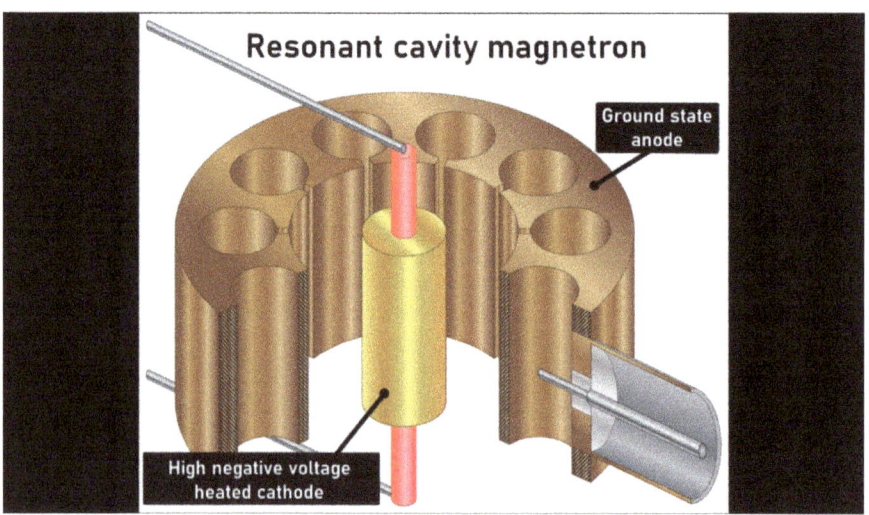

CHAPTER 8   43

Cavity magnetron

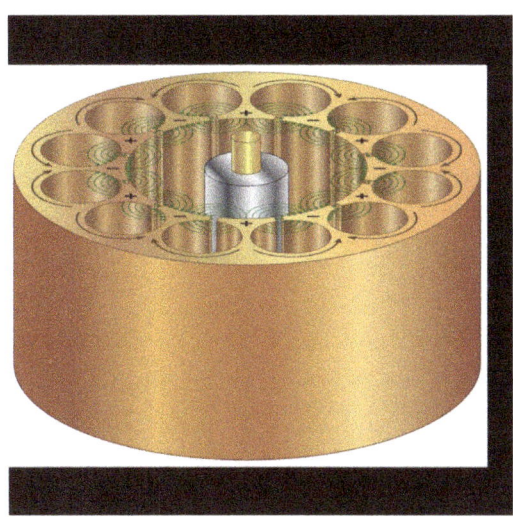 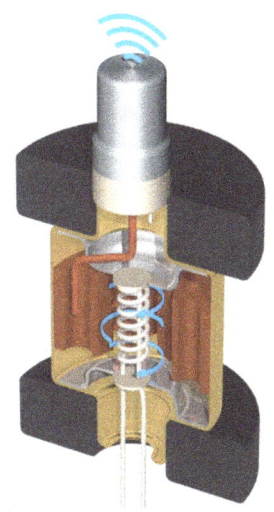

***

[1] www.wikipedia.com Christopher Dunn author 1998 "The Giza Power Plant"
  [2] www.wikipedia.com "X-ray tube"
  [3] www.wikipedia.com "Cavity magnetron"

Chapter 9

Iron plate and the Signal Beacon

During a combustion event, the hypothesis suggests that some of the unburned fuel gas is forced into and accelerated up the southern shaft, the shockwave induces a high voltage electromagnetic pulse that ionizes the hydrogen gas while the expansion of steam causes a sharp rise in combustion chamber pressure. The ionized gas is propelled and accelerated up the shaft, producing long wave electromagnetic energy and a blip of shorter wavelength radiation when some of the high velocity electrons collide with the ground state anode Iron plate located at the top of the shaft.

The Hydrogen ions { Hydrogen atoms that have been stripped of their electron } are positively charged bare protons since the Hydrogen atom does not contain a neutron, except in the case of Deuterium or Tritium and would be attracted to the grounded Iron plate. The absorption of the positively charged protons by the Iron plate would cause a momentary bias or availability of charge carriers causing the plate to become weakly attractive to the high velocity electrons.

CHAPTER 9

Iron plate

In the spring of 1837, a team of explorers led by Colonel Howard Vyse were conducting exploratory blasting on the Great Pyramid. On May 26, 1837 the remains of an Iron plate was discovered by J. R. Hill after blasting the outer two tiers of stone from the mouth of the King's chamber's southern shaft.[1] The remains of the plate were studied and found to be wrought Iron and measured 26 cm long, by 8.6 cm wide and .4 cm thick. The remains were also determined to be of ancient origin due to the composition of the corrosion on its surface.

***

[1] www.youtube.com LINES IN SAND 2023 "Investigating the Iron plate found in the Great Pyramid"

Southern shaft and the Iron plate

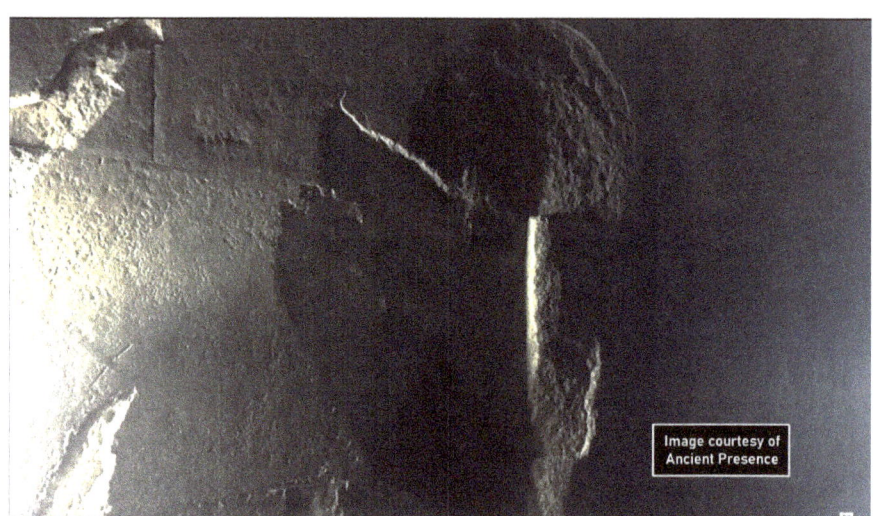

Chapter 10
Wireless receivers, user devices and the Wear & Tear evidence

Wireless receivers and user devices

While we're left with plenty of room to speculate about the purpose and function of the massive obelisks from the ancient world, the prime example we have of a wireless receiver and user device is the "Dendera Lamp" relief carvings in the Temple of Hathor. The artwork is of course an artistic impression that's open to interpretation by anyone who sees it. Here the hypothesis suggests that a portion of the artwork depicts a man holding an ampule of ionizing gas near a wireless receiver, the tall structure with a stack of four disks atop the column known as a "Djed Pillar". It's thought that the receiver attenuates the wireless power and yields a high voltage.

It's well known that holding an ampule of ionizing gas near a high voltage source such as a Tesla Coil will cause the gas within the ampule to glow.[1]

In the artwork, the ampule that the man is holding appears to feature a cable connected to a box. The hypothesis suggests that the box contains a capacitor and that during operation, a high voltage and small current is exchanged between the receiver and the capacitor downstream of the ampule. The small current moving through the ampule at four hertz causes an arc to propagate and extinguish four times a second, leading to the artistic rendering of a wiggly snake within the ampule.

***

[1] www.youtube.com Thoisoi 2 2023 "Hydrogen, the lightest gas in the universe"

CHAPTER 10

Gas ampules

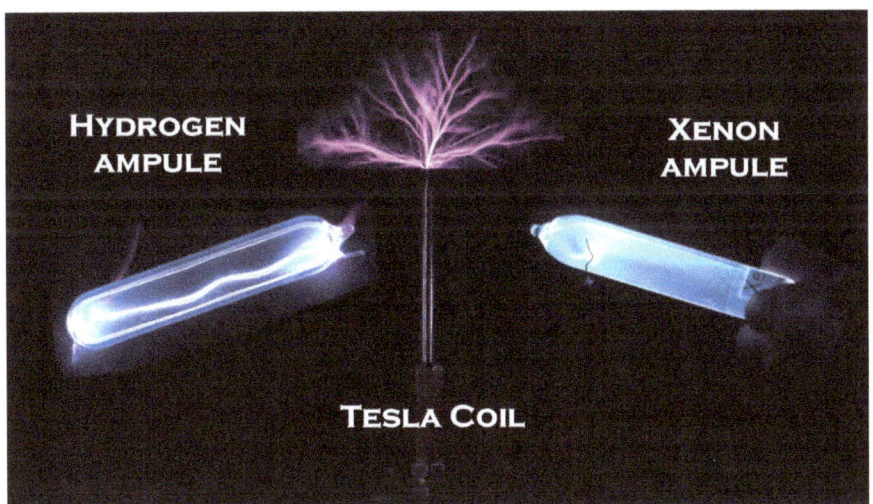

The Wear & Tear evidence

The limestone Great Step

The substantial erosion of the limestone "Great Step" is of particular interest since it shows two distinct erosion patterns. The pure water created within the combustion chamber could be considered Hydric acid due to having a low PH and is corrosive to limestone, which also provides some clue as to the lifespan of the system's operation. The erosion appears to show a spray or blast pattern from the Antechamber plus a trickle down the middle where the erosion reaches a bit less than a meter in depth on the leading edge of the stone.

While the dissolved limestone would serve to buffer the PH of the water in the system, the quantity of dissolved Calcium carbonate could help to account for the Queen's chamber being found encrusted with some type of salt deposits.

The Antechamber

The King's chamber's Antechamber bears an appearance of being scrubbed with hot, high velocity gasses somewhat similar to the interior surface of an engine's exhaust manifold.

CHAPTER 10

***

The King's chamber

The condition of the sarcophagus is of particular interest since a close inspection of the stone box shows extensive erosion all around the top of the box, consistent with damage from a repeated force causing micro erosions occurring over a period of time. The King's chamber's south wall also appears to show substantial impact damage around the southern shaft.

King's chamber south wall

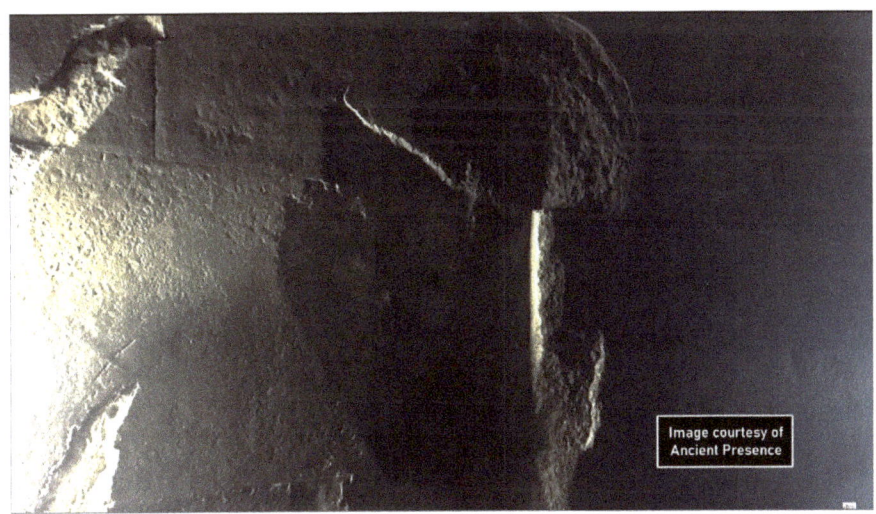

King's chamber Wear and Tear

CHAPTER 10

The Great Step circa 1910

This appearance of wear and tear evidence together with the aspects of physics, makes sense of the otherwise bizarre presence and physical relationships of the building's features and gives rise to the *"Internal combustion, piezoelectric, electrochemical hypothesis of the Great Pyramid of Giza"*.

While Occam's razor might suggest that the simplest answer is probably true, the conventional narrative of Dynastic Egyptian tomb for the the Great Pyramid is contraindicated. We know that Dynastic Egyptian tombs were usually concealed to hide their location from potential grave robbers. We also know that royal tombs featured paintings and carvings that tell about the life and accomplishments of the person entombed there.

Summary

The traditional narrative of the past was born of people from centuries ago with little to no knowledge about the concepts of physics that we take for granted today.

CHAPTER 10

Viewed through the eyes of modern knowledge, the Great Pyramid presents itself as a machine, a remnant of high technology from an advanced civilization lost to history a long time ago. While the Great Pyramid presents a solid state solution with no moving parts, the configuration and shape are proprietary since the operating strategy is dependent on gravity. One has to wonder if we can bring this old world tech into the new world and what it might look like using modern materials and some creative configuration. Considering the recent interest in using Hydrogen as a fuel or moving to a Hydrogen economy like the Empire of Japan, perhaps an adaptation of this technology to produce a large current and high negative voltage could present a useful power supply solution in the advancement of H^2 fuel production.

Inscription found at the entrance of the Great Pyramid

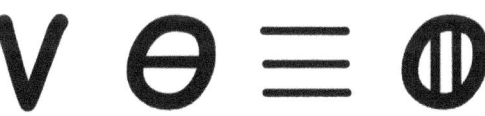

Masuline force granting feminine

Chapter 11

Volumes and values

While quantifying the energy relationships of the system seems daunting at best, the things we know about physics provide a framework to an understanding of the Great Pyramid. For example, we know that it takes 2.3 kilowatts of heat to drive the phase change of one liter of water from a liquid to a gas, producing 1.65 cubic meters of steam. We also know that it takes 13.2 megajoules, or 3,670 watts of electrical energy for one hour to drive the electrolysis of one liter of water to produce 1.2 cubic meters of Hydrogen gas and 0.6 cubic meters of Oxygen. We know the pyrolysis temperature of water to be about 3000^0C or 5400^0F and the temperature of high voltage electrical arcs exceeds the decomposition temperature of water.

The interior dimensions of the cavity spaces within the structure are well known and the total volume of the interior cavities works out to somewhere between 2000 - 2200 cubic meters of space above the level of the drain. Below the level of the drain, the horizontal passage leading into the Queen's chamber steps down a full half meter.

The depth of the step combined with the area of the floor gives the Queen's chamber a liquid water holding capacity if about 16 cubic meters or 16,000 liters of water. Opposite the Queen's chamber, the water reservoir has an estimated capacity of about 65 cubic meters or 65,000 liters of water. The drain cavity and plumbing are thought to serve as a drain path and act as a buffer or silencer during operation.

The other interior volume

Equally important and difficult to estimate is the Hydrogen gas capacitance of the building's stonework. Particularly in the upper portion of the building and around the Queen's chamber air shafts, the cracks and gaps between the stones where the H^2 gas can leak and be displaced into on its path upward.

If the low Magnesium fine Tura limestone casing were fit "perfectly" to create a gas tight seal, the H^2 gas would be retained as it rises to the highest points available within the structure. This Hydrogen gas capacitance presents a key aspect for the system's operation by retaining the gaseous dielectric medium to serve as an electrical conductor at high voltages through primary ionization and act as an amplifier at increasingly higher voltages through the avalanche effect of secondary ionization as described in Townsend's theorem.

Chapter 12

Conclusion

It seems pretty easy to see that the cycling of about two cubic meters of water to steam and back again to a liquid, would be sufficient to keep the internal cavities filled with steam and saturated with condensate as the condensed water recycles back to the steam chamber.

The volume of Hydrogen gas needed to saturate the stonework and fill the gas cavities in unclear, but thought to be about the same as the proper internal volume of the structure. This suggests that about four cubic meters of water, or 25% of the Queen's chamber water supply would be needed to achieve start up and system operation. Once the system starts up and begins running, each combustion event would send a pressure wave that travels the length of the Grand Gallery and reflects off the water in the reservoir. The impact of the pressure wave would cause water to splash from the reservoir and into the Queen's chamber horizontal passage, keeping the Queen's chamber filled with water during system operation.

It seems only a matter of time until water becomes the fuel stock of choice for clean, sustainable energy once our technology catches up and finds its niche to supply our energy needs for tomorrow.

With this promise of the future in mind, I'll close with the following thought. The only equation suggested in this book is *"Townsend's equation for current growth in a gaseous dielectric medium"* and the only formula presented is the formula for water. **H^2 plus O equals water and the highest specific impulse of all known fuels.**   Terry E. Lee

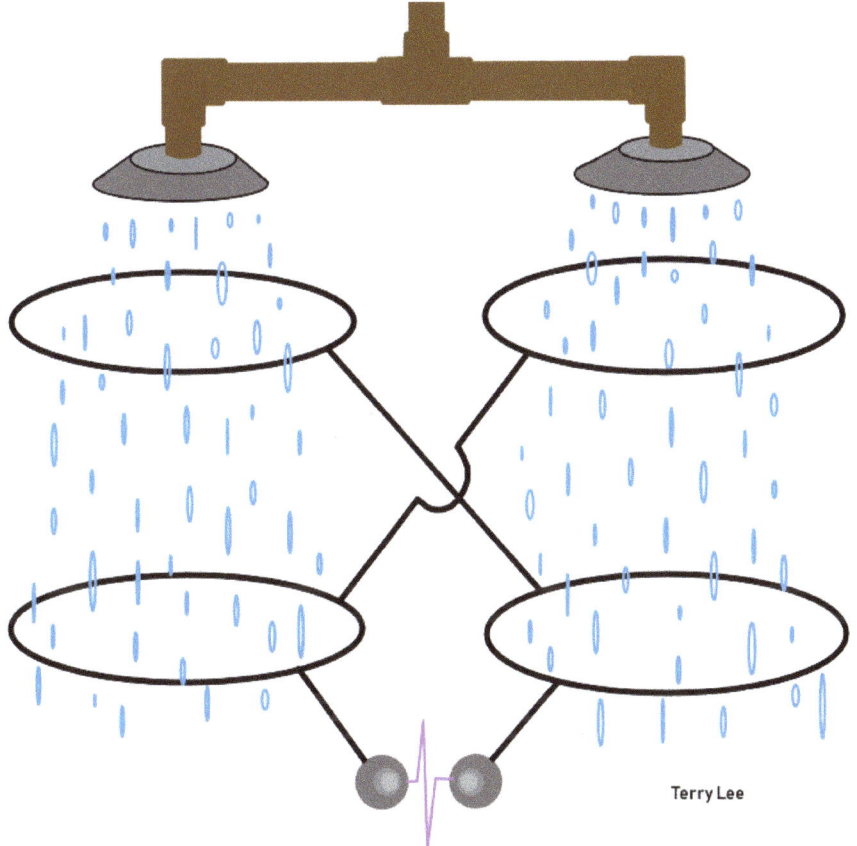

Kelvin's Water Dropper experiment

Also known as Kelvin's Rainstorm, or Lord Kelvin's Thunderstorm Generator. The experiment shows that dribbling water droplets through a set of cross connected coils will produce high voltage electrical arcs of 10kv - 20kv between the electrodes.

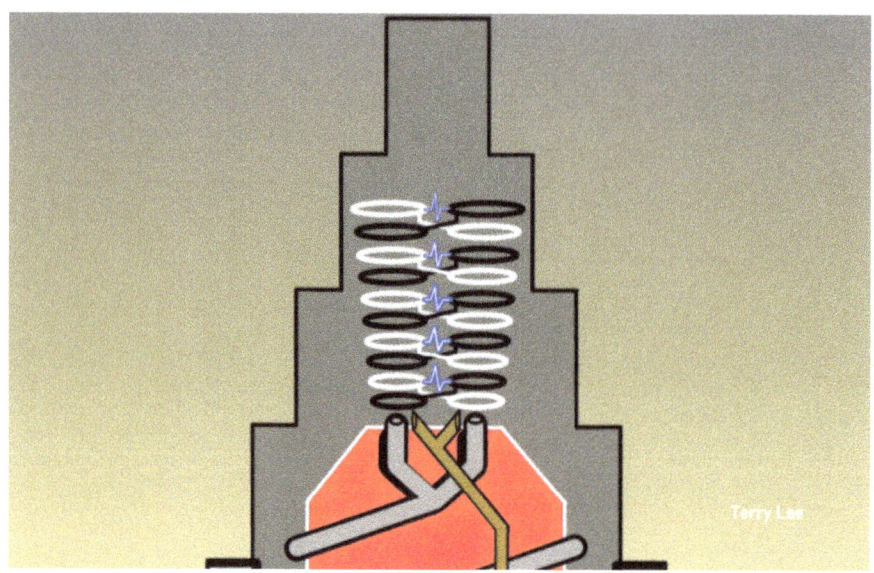

CHAPTER 12

Impression of operation

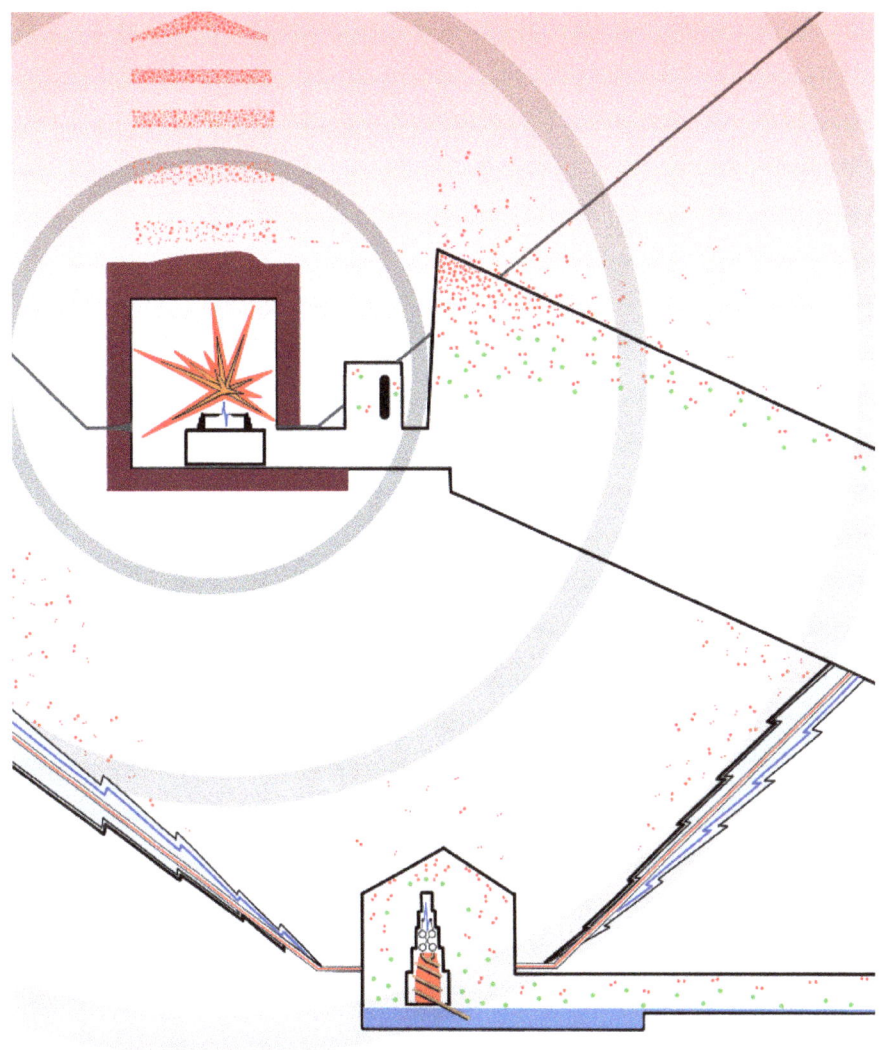

Terry E. Lee

Print index

Index	Page
Air shafts	2, 57
Amplifier	13, 28, 37, 57
Antechamber	2, 17, 50
Avalanche effect	28, 31, 33, 36, 57
Beta decay	20
Calcium carbonate	50
Combustion chamber	17, 19, 21, 22, 24, 33, 36, 44, 50
Conductors	2, 30
Copper	30
Cusco, Peru	5, 71
Dendera Lamp	47
Djed Pillar	47, 49
Drain	13, 56, 69

Print index

Electrodes	2, 15, 34, 59
Electrolysis	2, 13, 30, 33, 56
Electromagnetic pulse	21, 31, 36, 38, 44
Electrons	20, 28, 31, 33, 36, 41, 44
EMP	30, 31, 36
Erosion	2, 50, 51
Exothermic reaction	17
Expansion of steam	17, 44
Frequency	20, 33, 39
Frequency of operation	18, 22
Fuel metering	17
Gas cavity Amplifier	37
Gas purge	3, 16
Gaseous dielectric medium	13, 28, 30, 36, 57
Giza Power Plant	41
Grand Gallery	17, 18, 22, 58
Great Step	2, 50, 53

Print index

H² gas	16, 30, 36, 55, 57
Heat source	2, 10, 13
HHO fuel	16, 17, 33
High voltage	2, 13, 21, 28, 30, 38, 41, 44, 47, 48, 56
Howard Vyse	45
Humidity	16, 22, 33
Hydrogen	2, 13, 16, 21, 29, 44, 55, 56, 57, 58
Hydric acid	50
Igniter	2, 19, 20
Internal combustion	2, 18
Ionization	13, 28, 29, 30
Ions	28, 33, 44
Iron plate	44, 45, 46
J. R. Hill	45
Jedd Pillar {Djed Pillar}	47, 49
Kelvin's Water Dropper	15, 59
King's chamber	2, 33, 41, 45, 50, 51, 52

Print index

Lord Kelvin's Thunderstorm	13, 15, 59
Machu Picchu	5
Magnetron	41, 42, 43
Nasa	11
Negative voltage	41, 55
Niche	2, 10, 11, 13
Nikola Tesla	1, 38
Occam's Razor	54
Ollantaytambo	5
Oxygen	2, 13, 16, 21, 24, 56
PH	50
Piezoelectric effect	24, 27
Plasma HHO production	33
Power output	36
Pressure pulse	22
Pressure wave	18, 19, 58
Primary ionization	13, 28, 57
Protons	44
Pyrolysis	2, 13, 33, 56

Print index

Quartz crystal	21, 24, 25
Queen's chamber	2, 10, 11, 33, 35, 56, 57, 58
Radiation	41, 44
Radioisotope	20
Relieving chambers	36
Resonance	33, 38
Sacsayhuaman	5, 71, 72, 73
Salt deposits	2, 50
Sarcophagus	2, 51, 52
Secondary ionization	13, 28, 31, 33, 36, 57
Shockwave	17, 24, 44
Signal beacon	44
Silicon / Oxygen lattice	24, 26
Snap reactor	11
Spark gap	20, 38
Specific impulse	17, 59
Speed of sound	22, 33
Steam	13, 16, 17, 22, 33, 44, 56, 58
Steam chamber	22, 30, 32, 33, 58
Stinger exhaust	18
Subterranean chamber	11

Print index

Temple of Hathor	47
Tesla Coil	38, 39, 40, 47
Thermal signature	2, 50
Townsend's theorem	28, 33, 57, 59
Tuning chamber	18, 22
Voltage source	21, 24, 31, 36
Volumes and values	56
Wardenclyffe Tower	38, 40
Water fill	3, 10, 11
Water reservoir	11, 58
Water to fuel converter	32
Water vapor	16, 17
Wear and tear evidence	2, 47, 50, 51, 52, 53, 54
Wireless receivers	47
Younger Dryas	3

Chapter 13
Bonus pics

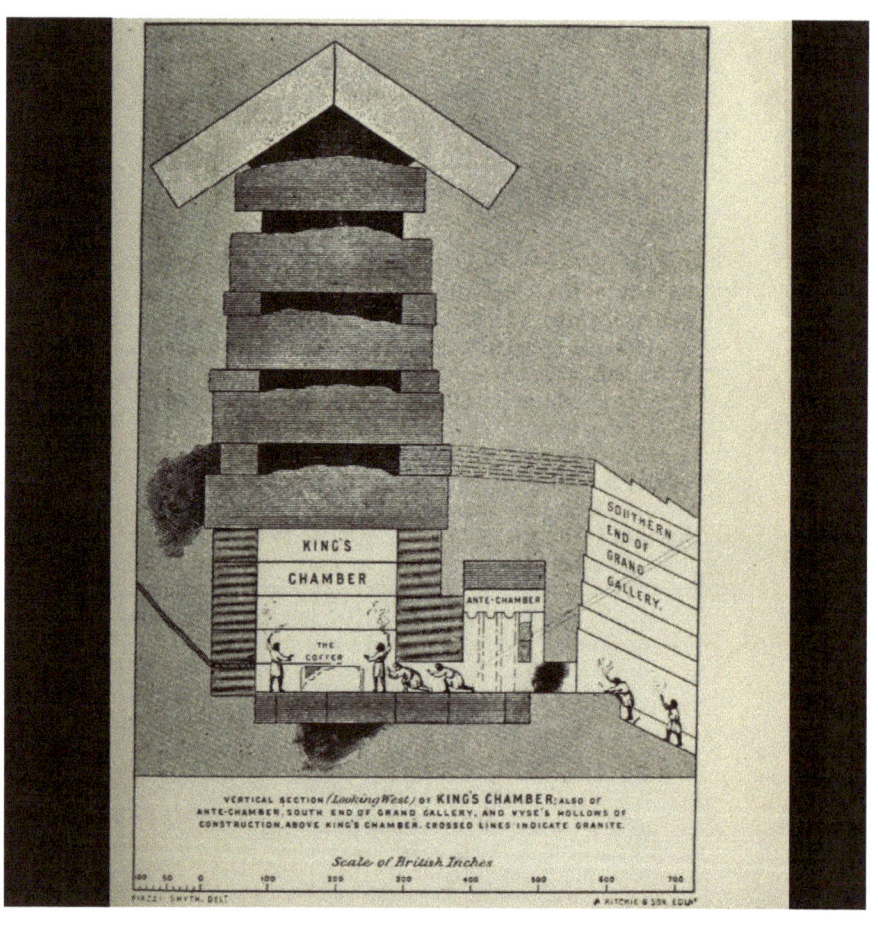

CHAPTER 13

Broken drain circa 1900

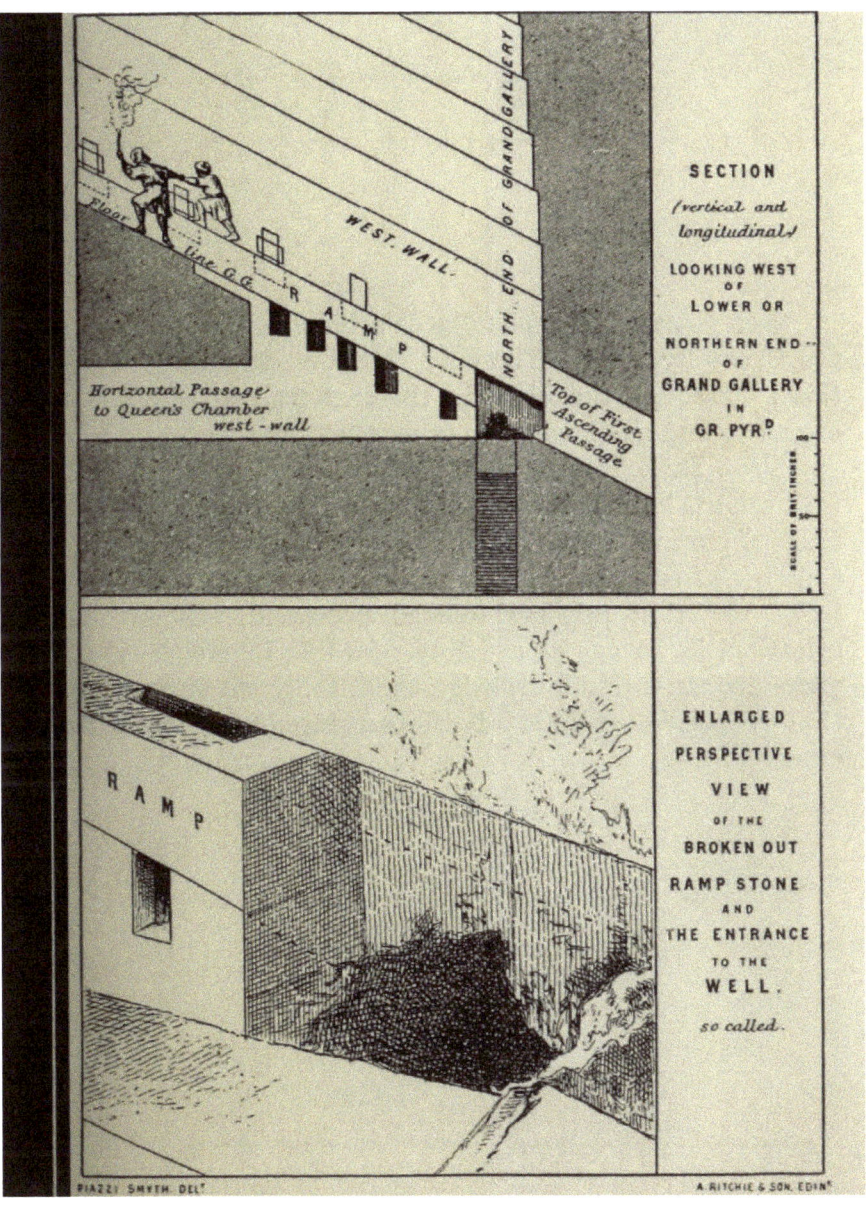

Karnak, Egypt

Axum, Ethiopia

Cusco, Peru

Sacsayhuaman near Cusco, Peru

Hill top at Sacsayhuaman

Suspected resonant cavity device

Close up